MARTIN THE MUSHROOM

THE TREASURE ADVENTURES™ SERIES

STORY BY **E. BERGLARK**

ILLUSTRATIONS BY
CHERYL CROUTHAMEL

RED
ROVER
press™

RED ROVER press

A DIVISION OF THE PUBLISHING CIRCLE

For permission requests, write to the publisher, addressed
"Attention: Permissions Coordinator,"
at the address below.

admin@ThePublishingCircle.com
or
RED ROVER PRESS
Regarding: E. Berglark
19215 SE 34th Street
Suite 106-347
Camas, Washington 98607

MARTIN THE MUSHROOM / E. BERGLARK
ISBN 978-1-947398-64-1 softcover
ISBN 978-1-947398-67-2 hardcover

DEDICATION:

Eric & Alexa Berger & Ryan & Laura Clark
would like to dedicate this book to
Phoenix, Weslee, Leo & Minka.
And last but not least,
YOU.

After all, we are all made of stardust.

This story takes place 167 million years ago in a time before humans known as the Jurassic Period.

On a quiet mountain overlooking a beautiful jungle, a Mushroom family lives inside a cave. They have just welcomed a new baby mushroom, called a "spore", into the world. His name is Martin the Mushroom.

Martin is a curious little mushroom.

"Mama, Papa, what are all those fuzzy things growing out of the ground?" Martin asked as he grew older.

"Those are called ferns, Martin," his parents replied.

"What do they do?"

"They release oxygen into the air and are food for dinosaurs."

"You mean those big creatures we see out there?"

"Yes, those creatures are called dinosaurs," his parents answered.

Martin pretended he was
outside his cave exploring
the mountains he could see
in the distance.

"Hey you!" Martin heard from above.

Martin was afraid of the dark and didn't want to turn around, but when he did, he saw a beautiful cluster of shiny purple rocks.

"H-h-hello?" Martin nervously replied.

"Why don't you quit yapping all day!" the shiny rock said.

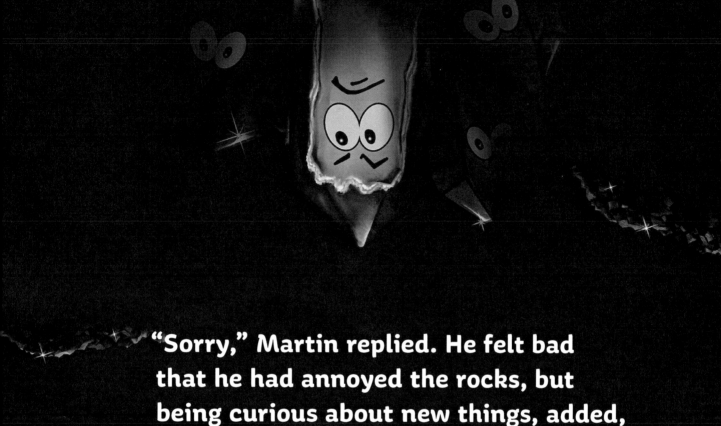

"Sorry," Martin replied. He felt bad that he had annoyed the rocks, but being curious about new things, added, "My name is Martin. What's yours?"

"My name is Geo," the rock grumbled, "and I'm a Geode. My family and I have lived in this cave long before yours!"

"Nice to meet you, Geo. Sorry I bothered you, but I love exploring!"

Martin's grandfather turned to him and said, "Don't mind them, Martin. They have been grumpy as long as I can remember."

"But why, Grandpa?" asked Martin.

"They are jealous because they're stuck on top of the cave and don't have the beautiful views we have," Grandpa said.

"Hi, Geo!" yelled Martin. "Do you want to be friends?"

"Leave me alone," Geo replied gruffly.

Martin felt sad that Geo didn't want to be his friend, but didn't know what else to do.

As the sun rose, Martin woke up to something tickling his roots.

"What is that?" Martin asked his family, laughing.

"That's an earthquake, Martin. It's when the ground shakes and tickles your roots. See those rocks falling down the mountains over there?"

Suddenly, there was a large BOOM! Then black smoke began shooting out of the mountain.

"WWWOOOWWW!" Martin yelled.

"That's no earthquake," said his grandpa, "that's a VOLCANO!"

"MARTIN! MARTIN!"
yelled Geo in a
frightened voice.

"What's up, Geo?" Martin
thought Geo looked really scared.

"If you tell me what you see, I promise
I won't be mean to you anymore."

This was Martin's chance to finally bring peace to the families once and for all and to make a new friend!

"Okay, Geo. See that pointy rock near you? Imagine that, but a million times bigger, and the very top is shooting out black clouds."

Geo closed his eyes and tried to imagine what Martin saw.

Geo asked his family, "How are you guys so calm?"

"We have been here for thousands of years, Geo," said his father, "so we have experienced this a few times. Even though we don't know what it is, we have heard it and felt it before."

"Are you listening to Martin?" asked Geo. "He is describing it to us."

"Martin, do you see that bright red liquid that moves like water? That's lava!" said Grandpa Mushroom.

"Yes, I see it," Martin replied.

"Eventually it will dry and turn green with new plant life until the volcano erupts again."

As time passed, the cave became alive with communication.

"Hey Martin, in the past we have seen dinosaurs gobble up mushrooms just like you. Not all dinosaurs just eat meat you know," said Father Geode.

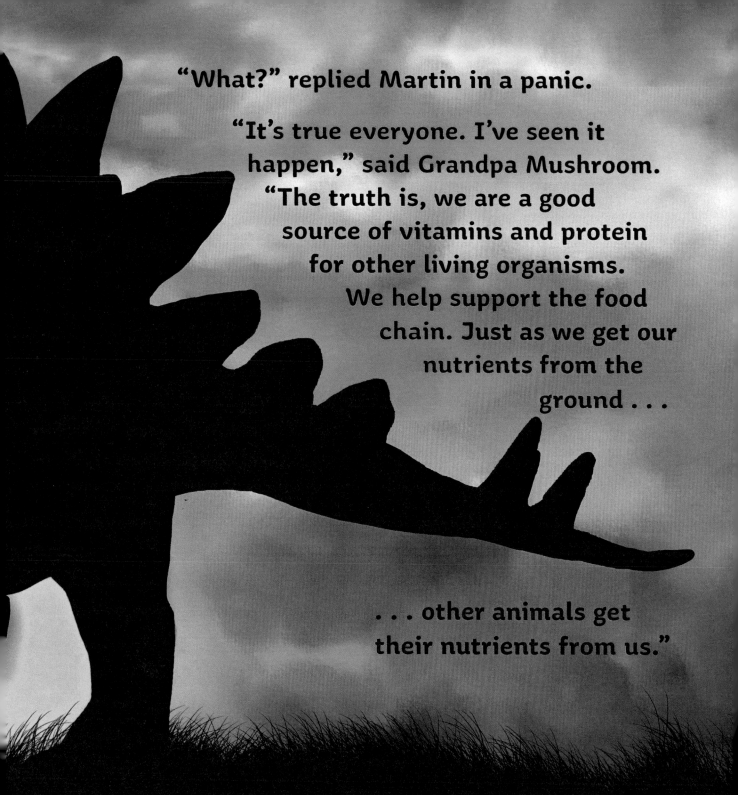

"What?" replied Martin in a panic.

"It's true everyone. I've seen it happen," said Grandpa Mushroom. "The truth is, we are a good source of vitamins and protein for other living organisms. We help support the food chain. Just as we get our nutrients from the ground . . .

. . . other animals get their nutrients from us."

"But what will happen to us then?" Martin asked.

"We turn into poop and sink into the ground to have a chance to grow again," said Grandpa Mushroom.

"PPHHEEWWW!" said Martin. "Let's hope *that* never happens! I never want to leave you, Geo."

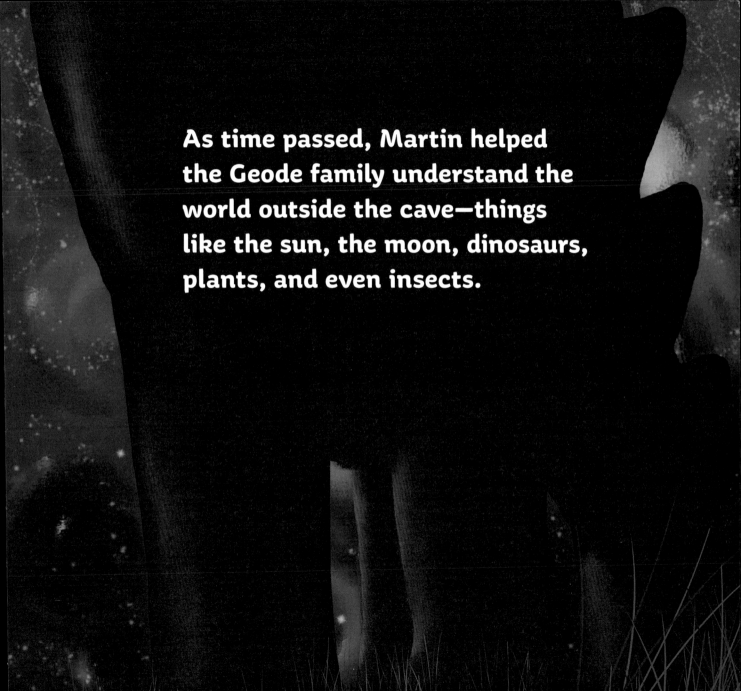

As time passed, Martin helped the Geode family understand the world outside the cave—things like the sun, the moon, dinosaurs, plants, and even insects.

"GEO!" shouted Martin.

"Martin!" yelled Geo. "What's that shaking?"

"It's a stegosaurus!" hollered Grandpa Mushroom.

"This is it, Geo," Martin said calmly. "It's our time to become part of the food chain. I'll never forget our friendship."

"I don't want you to leave!" pleaded Geo.

"Don't be sad, Geo. This is why we are here, and we are going to fulfill our destiny," Martin said.

"I love you, Martin. You are my best friend!" Geo yelled.

"I love you, too, Geo. You are *my* best friend!"

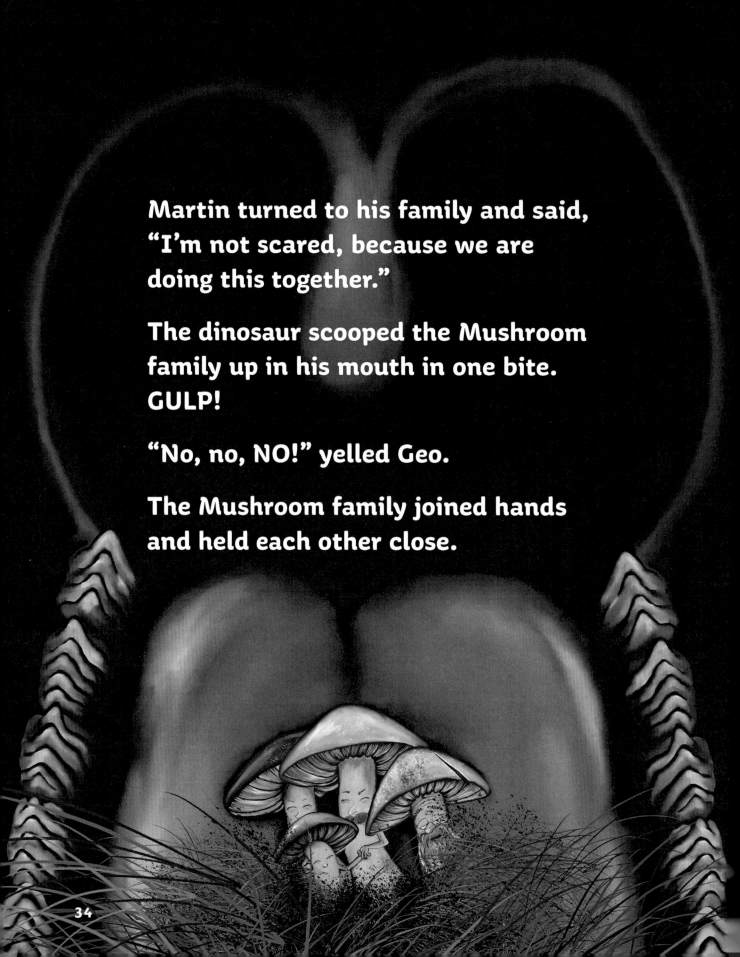

Martin turned to his family and said, "I'm not scared, because we are doing this together."

The dinosaur scooped the Mushroom family up in his mouth in one bite. GULP!

"No, no, NO!" yelled Geo.

The Mushroom family joined hands and held each other close.

"*Wh . . .* where am I?" asked Martin.

"Martin! I can't believe it! You're back!" yelled Geo.

"Hooray!" cried Martin. "The dinosaur's stomach must have protected us somehow. I can't believe it!"

Martin and the Mushroom family were pooped out in the same place they had always lived, and once again could gaze out at their beautiful valley.

Throughout the centuries, the Mushrooms described to Geo and his family what they saw.

Many things changed over the years with different types of trees, flowers, plants and animals changing and adapting in order to survive.

During these millions of years, many different visitors came into the cave—bears, mountain lions, wolves, mice, bats, and even humans.

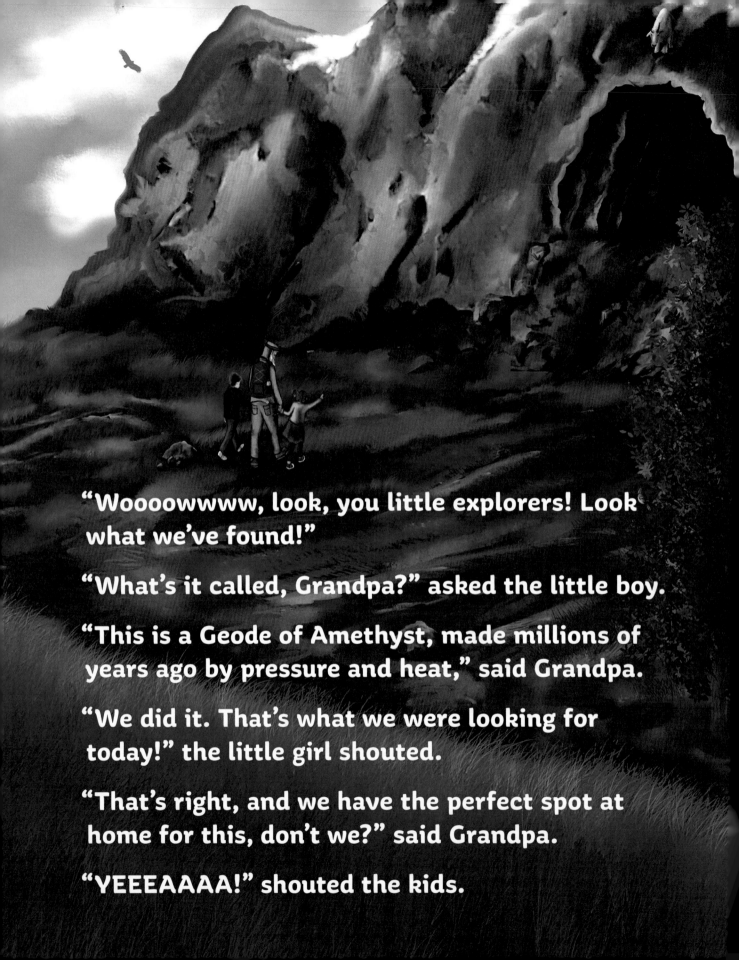

"Woooowwww, look, you little explorers! Look what we've found!"

"What's it called, Grandpa?" asked the little boy.

"This is a Geode of Amethyst, made millions of years ago by pressure and heat," said Grandpa.

"We did it. That's what we were looking for today!" the little girl shouted.

"That's right, and we have the perfect spot at home for this, don't we?" said Grandpa.

"YEEEAAAA!" shouted the kids.

"Martin! Did you hear that? They know my name!"

"I heard, Geo. How awesome is that? I told you that you were something special," said Martin.

Geo could not stop smiling.

Grandpa took out a chisel and hammer and started tapping around Geo.

"Ha-ha-ha-ha. That tickles!" Geo giggled.

"Are you guys okay?" asked the Mushroom family.

"Never better!" said the Geodes.

After a few taps and a THUMP the whole Geode family was in Grandpa's hands. The Geodes had never been out of their dark cave and had to squint their eyes when sunlight touched them for the first time.

"Wow!" the kids said as they gathered around their Grandpa to get a better look at the Geodes.

"Ready to head home?" Grandpa asked.

"I can hardly wait to show everyone what we found," the little girl said.

As they were walking back past the Mushroom family, the little girl tripped over them and fell down.

"Are you okay?" asked Grandpa.

"Yeah, I'm fine. I just tripped on this rock," said the little girl.

"Let me see that," said Grandpa. After he studied the rock, he started to laugh. He scratched his head.

"Well, I'll be," he said. "This is coprolite,"

"*Corpo*-whaaaaaaat?" asked the little boy.

"Cop-ro-lite. It's fossilized dinosaur poop! This means millions of years ago a dinosaur came into this very cave and went POOP! Then minerals transformed this poop into coprolite."

"EWWW! Ha ha ha ha! That doesn't look like poop," said the little boy.

"That's really cool, Grandpa. I can imagine it now," said the girl.

"We're not *really* going to take this home, are we?" asked the boy.

"You bet we are! This just as special as the geode we took down. This is worth a fortune!"

The children looked at each other in amazement.

And just like that, Martin and his family were carefully placed into the same sack as the Geodes.

**The Mushrooms and Geodes
studied each other. They had never
been so close to each other before.**

After a long hike, the Geode and Mushroom families were carefully lifted out of the bag and place inside a velvet-lined box. The man then set them on a table next to a window.

The room was beautiful and filled with light. Other treasures of different shapes, sizes, and colors surrounded them. Everyone had a spectacular view.

Geo could see all the things he had only heard about.

"Well, Martin," Geo sighed in delight, "I guess it's true. We are all special in our own way!"

"Howdy, y'all!" interrupted a voice beside them. "You folks look a little tuckered out from your journey. Sure would love to hear your story . . . then I'll tell ya mine."

GLOSSARY

Amethyst: Amethyst is a violet, or sometimes purple, variety of quartz.

Centuries: A century is 100 years. The plural of century is centuries.

Friend: A person you like and who likes you back. True friends support each other.

Geode: A crystal surprise that lives inside a rock. The rarest and most valuable geodes contain amethyst crystals the same color as Geo and his family.

Jurassic Period: A period in time about 201 million years ago that lasted for 56 million years.

Lava: Molten rock that flows out of an erupting volcano.

Peace: When life is calm and no one is fighting.

Stegosaurs: A small-headed dinosaur of the late Jurassic Period, with a spiked tail and a series of large triangular bony plates along its back. Stegosaurs only ate plants.

Vitamins: They are in our food and are essential for helping us grow and stay healthy.

Made in United States
North Haven, CT
24 June 2022

20595265R00033